An Introduction to Activated Carbon Adsorption Technologies

2nd Edition

J. Paul Guyer, P.E., R.A.

Editor

The Clubhouse Press
El Macero, California

CONTENTS

1. INTRODUCTION

**2. REGENERATION, REACTIVATION, AND DISPOSAL
OF SPENT ACTIVATED CARBON**

(This publication is adapted from the *Unified Facilities Criteria* and other resources of the United States government which are in the public domain, have been authorized for unlimited distribution, and are not copyrighted.)

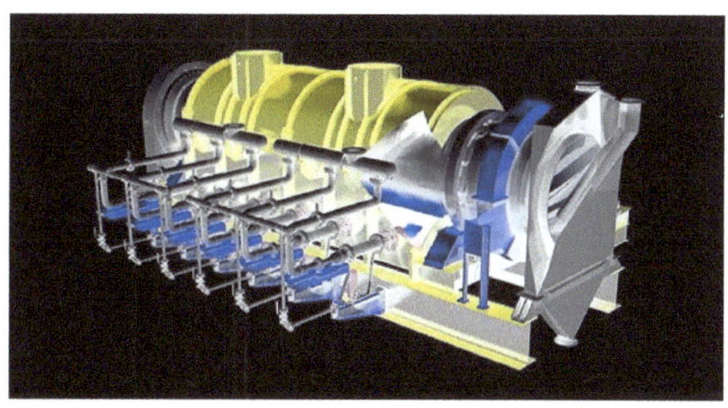

1. INTRODUCTION

1.1 PURPOSE. This discussion provides practical guidance for the design of liquid and vapor phase devices for the adsorption of organic chemicals. The adsorptive media addressed include granular activated carbon (GAC) and other alternative adsorption carbon media, such as powdered activated carbon (PAC) and non-carbon adsorbents.

1.1.1 CARBON ADSORPTION.

1.1.1.1 LIQUID PHASE CARBON.

1.1.1.1.1 APPLICATIONS. Some typical rules of thumb for types of compounds that are amenable to carbon adsorption are as follows:

- Larger molecules adsorb better than smaller molecules.
- Non-polar molecules adsorb better than polar molecules.
- Non-soluble or slightly soluble molecules adsorb better than highly soluble molecules.
- Based on the polarity or solubility, or both, of the molecule being adsorbed, pH may have an influence on the extent of adsorption.
- Temperature increases the rate of diffusion through the liquid to the adsorption sites, but since the adsorption process is exothermic, increases in temperature may reduce the degree of adsorption. This temperature effect is negligible in water treatment applications and ambient vapor phase applications.

1.1.1.1.2 CHEMICALS ADSORBED. The following are examples:

- Alcohols are poorly adsorbed, they are very soluble and highly polar.
- Aldehydes are highly polar, and as molecular weight increases, the polarity decreases, and adsorbability increases.
- Amines are similar in structure to ammonia (NH_3) except the nitrogen is bonded to an organic group. Adsorption is limited by polarity and solubility.

- Chlorinated aromatics, and chlorinated aliphatics are low-polarity and low-solubility compounds, which make them generally quite adsorbable.
- Glycols are water-soluble and not very adsorbable. Higher molecular weight organic compounds will generally be more adsorbable owing to adsorptive attraction relative to size.

1.1.1.1.3 TYPES OF CARBON. Activated carbon is a generic term for a variety of products that consist primarily of elemental carbon. Numerous raw materials can be used to produce carbons, such as coal, wood, and pitch, and agricultural products such as cotton gin waste and coconut shells. Materials most commonly used for liquid phase GAC include both bituminous and lignite coal, and coconut shells. (a) Bituminous GAC is the one most frequently used for treating low concentrations of low molecular weight organic contaminants in the aqueous phase. Bituminous coal will also have a more fully developed pore distribution, including "transport pores" that improve the rate of adsorption making it effective for water treatment. Bituminous GAC has a relatively large surface area, approximately 900 m2/g, and an apparent density of approximately 0.50 g/cm3 (30 lb/ft3). These carbons are usually harder than other types except coconut, and, therefore, are more abrasion resistant, and can be more vigorously backwashed without damage. (b) Lignite GAC generally has less total surface area than bituminous GAC. It is a less dense, slightly softer coal, has a higher percentage of meso (transitional) macro pores, and is used more for larger molecules. Therefore, it is used more in decolorizing applications. Lignite GAC has a surface area of approximately 650 m2/g and an apparent density of approximately 0.50 g/cm3 (25 lb/ft3). (c) Coconut-shell-based GAC generally has a larger surface area than coal-based GAC, and a very large percentage of micropores. Coconut-shell-based GAC has a surface area generally over 1000 m2/g and an apparent density of approximately 0.50 g/cm3 (30 lb/ ft3). Coconut shell based carbons may not have the more fully developed pore structure that coal-based carbons have, because their source is vegetative material. Consideration should be given to rate of adsorption effects in liquid treatment. It is used primarily in vapor-phase applications. Coconut shell- based carbon is slightly more expensive to produce than coal-based GAC, since only about 2% of the raw material is recoverable as GAC, versus 8–9% for coal-based carbons.

1.1.1.1.4 ISOTHERMS. Isotherms are discussed elsewhere.

1.1.1.1.5 PRESSURE DROP. Headloss in liquid phase applications varies significantly, depending on the piping configuration, carbon particle size, contact time, and surface loading-rate (generally expressed in liters per minute per square meter [gpm/ft^2]). Typical loading rates are 80–240 Lpm/m^2 (2–6 gpm/ft^2); occasionally, loadings up to 400 Lpm/m^2 (10 gpm/ ft^2) are used. Loadings greater than 240 Lpm/m2 (6 gpm/ft2) generally result in excessive headloss through a typical arrangement that has two pre-piped, skid-mounted vessels in series (140 kPa [20 psi] or more primarily from piping losses). In any case, the manufacturer's literature should be consulted regarding the headloss for a specific application.

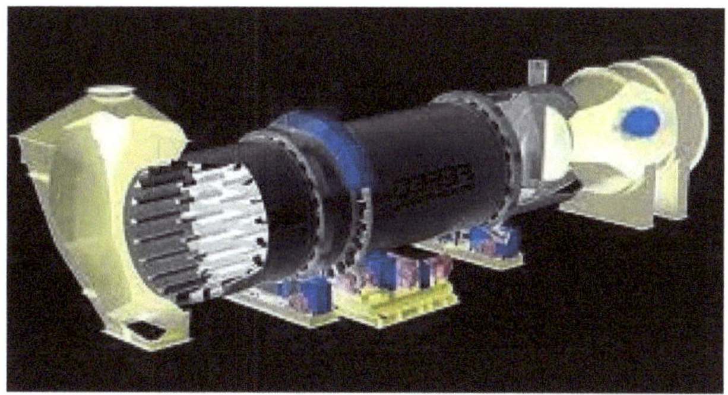

1.1.1.1.6 OPERATING PARAMETERS. (a) Contact Time. General rules of thumb for moderately adsorbable compounds such as TCE, PCE, and benzene are, first, to go from low ppm levels (approximately 1) to ppb levels requires a minimum empty bed contact time (EBCT) of approximately 15 minutes (some applications have shorter valid contact times given an effective process design), and, second, to go from a medium ppm range (approximately 10) to a low ppb range requires approximately 30 minutes EBCT. Some typical values are identified in Table 3-1. EBCT is related to the contactor dimensions as follows:

$$\text{EBCT} = \frac{V}{Q} \text{ or } \frac{LA}{Q}$$

where

V = bulk volume of GAC in contactor, m^3 (ft^3)
A = cross-sectional bed area, m^2 (ft^2)
L = bed depth, m (ft)
Q = volumetric flow rate, L/s (ft^3/min).

1.1.1.2 ADSORBER VOLUME. Once the optimum contact time (EBCT) and the carbon usage rate are established, the size (volume) of the adsorbers can be determined. Factors that affect the size of the adsorber include the change out rate as well as the carbon usage rate. Generally, for carbon contactor change out, you should consider schedules for other projects at an installation, as well as a reactivation company's fees, to determine the most cost-effective change out schedule. Typically, reactivation companies have compartmentalized trucks with a dry carbon capacity of 9100 kg (20,000 lb), which results in a saturated weight of 18,200 kg (40,000 lb), which is the load limit of most roadways. Off-the-shelf contactors range from 70 kg (150 lb) to as large as 9100 kg (20,000 lb). Optimum carbon usage should be based on column studies. The carbon usage rates at different contact times should be evaluated against the higher initial cost of the larger units and higher operation and maintenance costs of the smaller units. The carbon vessel should have an additional 20–50% bed expansion allowance built in for backwashing the carbon before you place the vessels in service. This expansion allowance is critical in systems where suspended solids are expected, or there is no pre-filtration. The adsorber volume is then calculated from:

$$V = \frac{(CUR \bullet COP)\ S.F.}{\rho}$$

Where:

V = volume of adsorber, ft^3

CUR = carbon usage rate, g/day (lb/day)

COP = carbon change out period, days

ρ = bulk density of carbon, g/cm^3 (lb/ft^3)

$S.F.$ = safety factor to provide extra non-carbon-containing volume for operational uncertainty, 1.2–2.5.

1.1.1.3 BED DEPTH. Bed depth is a direct function of the contactor diameter and volume. You can solve for the bed depth (L) knowing the adsorber volume (V) and adsorber bed area (A) using the equation:

- L = V/A

1.1.1.4 CARBON USAGE. Carbon usage can be estimated several ways. One method to estimate GAC usage is based on isotherm data using the relationships:

(1) For batch systems:

$$CUR = \frac{(C_o - C_e)\, F}{\left(\dfrac{x}{m}\right)_{C_o}}$$

(3-1)

(2) For flow through systems:

$$CUR = \frac{C_o\, V}{\left(\dfrac{x}{m}\right)_{C_o}}$$

(3-2)

Where

C_o = initial concentration (mg/L)

C_e = desired effluent concentration $\left(\dfrac{mg}{L}\right)\left(\dfrac{mg\ \ adsorbed}{g\ \ carbon}\right)$

$\left(\dfrac{x}{m}\right)_{C_o} = \dfrac{x}{m}$ value at concentration C_o $\left(\dfrac{mg\ \ contam}{g\ \ carbon}\right)$

$\left(\dfrac{x}{m}\right)_{C_e} = \dfrac{x}{m}$ value at concentration

$C_e = \dfrac{mg\ contamination}{g\ carbon}$

CUR = carbon usage rate (g/day)

F = volumetric flow rate of contaminated liquid treated/day (L/day).

Relationship 3-1 is generally used to estimate carbon usage for batch systems, and relationship 3- 2 is used for continuously operating flow through systems. For multiple constituent wastes, the constituents with the highest GAC usage rates, up to three, can be summed and the overall CUR estimated based on that sum. Estimates based on isothermal data will only provide a very rough estimate of GAC usage. In most cases a column test must be performed.

1.1.1.5 BACKWASHING. Backwashing is the process of reversing the flow through a media bed with enough velocity to dislodge any material caught in void spaces or attached to the media. Backwashing is essential before you bring a typical liquid phase downflow pressure column online. Backwashing removes carbon fines generated during the transfer from the shipping container to the contactors. Backwashing also helps naturally stratify the GAC bed, which reduces the likelihood of preferential channeling within the column, and, after future backwashes, helps keep spent carbon at the top of the bed. Redistribution of the adsorbent within a GAC bed that was improperly backwashed when initially installed could result in extending the mass transfer zone (MTZ), potentially reducing the overall adsorption capacity of the adsorber.

Backwashing a GAC bed prior to placing a new bed into service also helps de-aerate the bed, further reducing the potential for channeling. Periodic backwashing is usually recommended in the downflow adsorption systems most commonly used at HTRW sites, unless the water treated is low in dissolved and suspended solids. Periodic backwashing serves the same purposes that you would expect in any sand filtration system, to remove solids accumulation, reduce biological growth on the media, and reduce the headloss in the bed. The backwash rate will depend on the carbon density, particle size, and water temperature. Typically, a 30% bed expansion is accounted for in the design. This generally requires approximately 6.3–7.4 Lpm/m^2 (8–14 gpm/ft^2) at a water temperature of 13°C. The GAC manufacturer should be contacted to determine the optimum backwash rate for the carbon supplied. A portion of some poorly adsorbed constituents, such as carbon tetrachloride, may be desorbed during backwashing, but strongly held constituents are not affected.

Treating Groundwater for Non-Potable Use
Influent Concentrations at mg/L Levels, Effluent at the µg/L Levels

Example	Contaminant	Typical Influent Concentration (mg/L)	Typical Effluent Concentration (µg/L)	Surface Loading Rate (gpm/ft²)	Total Contact Time (minutes)	GAC Usage Rate (lb/1000 gal)	Operating Mode
1	Phenol	63	<1	1	201	5.8	Three Fixed Beds in Series
	Orthochlorophenol	100	<1				
2	Chloroform	3.4	<1	0.5	262	11.6	Two Fixed Beds in Series
	Carbon	135	<1				
	Tetrachloride	70	<1				
	Tetrachloroethylene						
3	Chloroform	0.8	<1	2.3	58	2.8	Two Fixed Beds in Series
	Carbon	10.0	<1				
	Tetrachloride	15.0	<1				
	Tetrachloroethylene						
4	Benzene	0.4	<1	1.21	112	1.9	Two Fixed Beds in Series
	Tetrachloroethylene	4.5	<1				
5	Chloroform	1.4	<1	1.6	41	1.15	Two Fixed Beds in Series
	Carbon	1.0	<1				
	Tetrachloride						
6	Trichloroethylene	3-8	<1	2.4	36	1.54	Two Fixed Beds in Series
	Xylene	0.2-0.5	<1				
	Isopropyl Alcohol	0.2	<10				
	Acetone	0.1	<10				
7	Di-Isopropyl Methyl Phosphonate	1.25	<50	2.2	30	0.7	Single Fixed Bed
	Dichloropentadiene	0.45	<10				
8	1,1,1 Trichloroethane	143	<1	4.5	15	0.4	Single Fixed Bed
	Trichloroethylene						
	Tetrachloroethylene						
9	Methyl T-Butyl Ether	30	<5	5.7	12	0.6	Two Single Fixed Beds
	Di-Isopropyl Ether	35	<1				
10	Chloroform	400	<100	2.5	26	1.2	Four Single Fixed Beds
	Trichloroethylene	10	<1				
11	Trichloroethylene	35	<1	3.3	21	0.2	Three Single Fixed Beds
	Tetrachloroethylene	170	<1				
12	1,1,1 Trichloroethane	70	<1	4.5	30	0.45	Two Fixed Beds in Series
	1,1 Dichloroethylene	10	<1				
13	1,1,1 Trichloroethane	25	<1	2	35	0.3	Single Fixed Bed
	Cis-1,1 Dichloroethylene	15	<1				
14	Trichloroethylene	50	<1	1.6	42	0.4	Two Single Fixed Beds
15	Cis-1,1	5	<1	1.9	70	0.25	Two Fixed Beds in Series
	Dichloroethylene	5	<1				
	Trichloroethylene	10	<1				
	Tetrachloroethylene						

Table 3-1

Example Case Studies

Treating Groundwater for Non-Potable Use Influent Concentrations at mg/L Levels, Effluent at the µg/L Levels

1.1.2 EQUIPMENT. Generally, steel pressure vessels containing granular activated carbon are used. In water treatment; steel vessels must have a protective internal lining to protect them from the corrosive effects of carbon in water. This lining should also possess good abrasion resistance to withstand movement of the hard carbon particles. The treatment systems range in capacity from 70 kg (150 lb) of carbon per unit to 9100 kg (20,000 lb) per unit. Under certain low-pressure applications, fiberglass or other plastic units may be used. In certain applications, ASME rated pressure vessels may be required. Units are generally skid-mounted, pre-assembled by the manufacturer, and delivered to the site. Larger units, i.e., 3 m (10 ft) in diameter, are difficult to ship pre-assembled, so major components, piping, and vessels are assembled in the field. Piping components are typically pressure-rated to match the vessels and included as part of the skid unit. A schematic presenting the major components is provided in Figure 3-1.

1.1.2.1 MOST LIQUID PHASE GRANULAR activated carbon systems are operated in series. This means passing all of the flow through one column bed, a lead column, and then passing flow through another similar sized column bed, the lag vessel. This method offers several advantages over a single column. The series configuration allows the maximum use of the GAC throughout the entire carbon vessel. This assumes, of course, that the mass transfer zone (MTZ) is contained within a single properly sized carbon unit. By placing two or more columns in series, the MTZ is allowed to pass completely through the first (lead) bed as the leading edge of the MTZ migrates into the second (lag) bed. By allowing this to happen, the maximum contaminant concentration is allowed to come into contact with adsorption sites in the lead vessel that require a greater concentration gradient (differential adsorption energy) to hold additional contamination. When the MTZ exits the lead vessel, that vessel is then exhausted, and requires change out with virgin or regenerated GAC. Even though the adsorption capacity of the lead vessel is exhausted, treatment continues in the lag vessel. Then, during change out, the lead vessel is taken off-line and the lag vessel is placed in the lead position. The former lead vessel is then replenished with GAC and then becomes the lag vessel and brought on-line.

1.1.2.2 A CRITICAL COMPONENT OF THE ADSORBER DESIGN is the underdrain (collection) system. This underdrain must be designed so that water is collected evenly, such that the mass transfer zone is drawn down in an even, or plug flow, manner to get full value from the installed carbon. In addition, the underdrain may also be used to introduce backwash water, and, therefore, it should be able to introduce water evenly across the entire bed cross section.

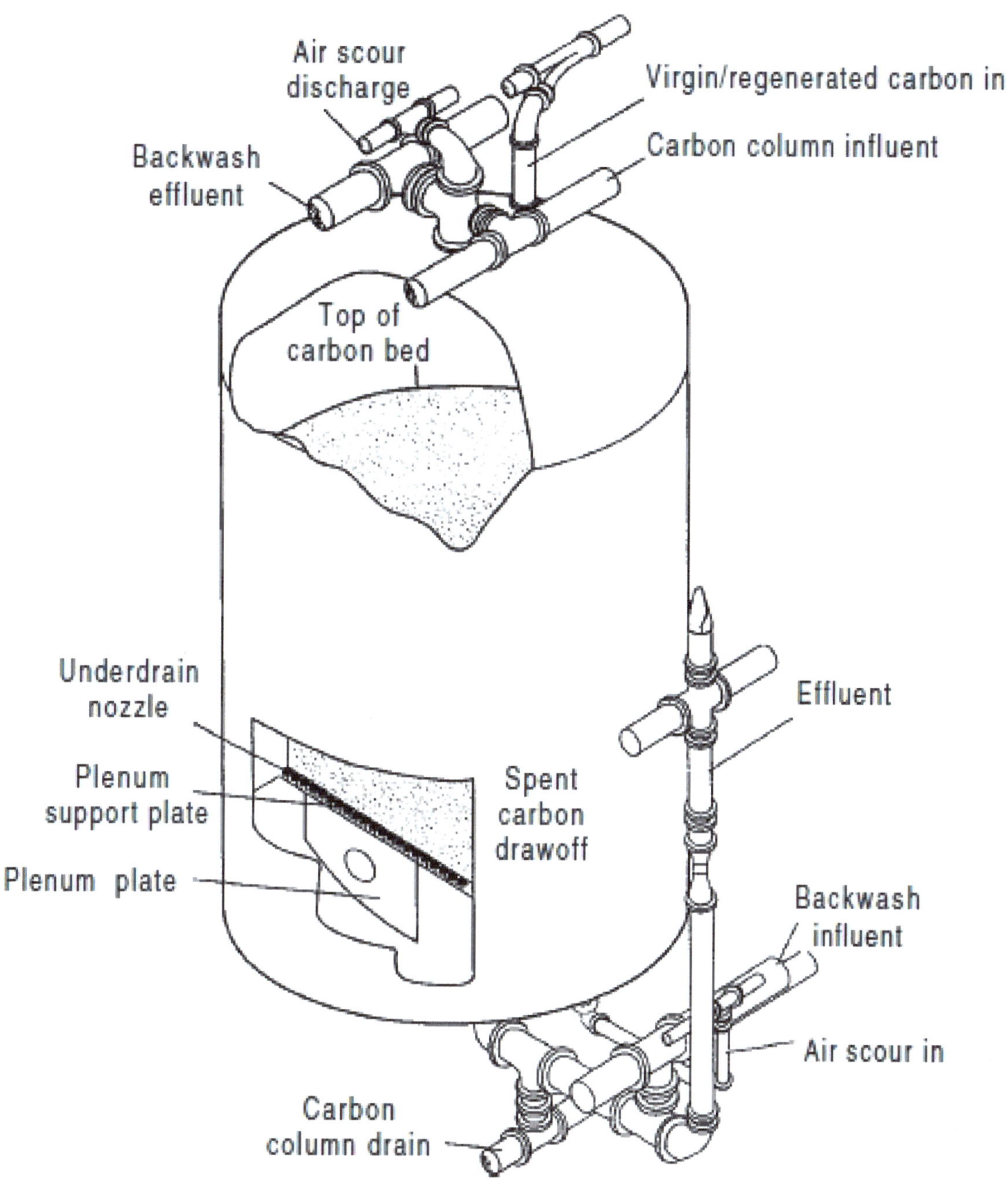

Figure 3-1

Schematic of carbon contactor.

1.2 VAPOR PHASE CARBON ADSORPTION.

1.2.1 APPLICATIONS. Vapor phase activated carbon adsorption is used to treat vapor emissions from processes such as air stripping, soil vapor extraction (illustrated in

Figure 3-2), thermal desorption, landfill off-gas, treatment process vessels, storage tanks, treatment buildings, and treatment processes (odor control).

1.2.2 CHEMICALS ADSORBED. Many volatile organic chemicals can be removed from vapor streams with activated carbon. In general non-polar organic chemicals adsorb better than polar organic chemicals, and higher molecular weight organic chemicals adsorb better than low molecular weight organic chemicals. Examples of hazardous waste chemicals that are easily adsorbed are chlorinated solvents such as trichloroethylene (TCE) and fuel components such as benzene, ethylbenzene, toluene, and xylene (BETX). Examples of chemicals that are not as easily adsorbed are aldehydes, ketones, and alcohols, although these do have better adsorptive characteristics than they do in the liquid phase, as solubility in water is not a factor, and adsorption will improve with increasing molecular weight. Most adsorption of VOCs by activated carbon is exothermic. The heat of adsorption is especially high with ketones, such as methyl ethyl ketone (MEK), and aldehydes. Heat from the vapor phase adsorption of these contaminants has actually built up and ignited bed fires in some installations (Shelly, 1994). The temperature of the bed should be monitored to prevent a "hot spot" from igniting a bed fire. Internal sprinklers are often installed in the carbon vessel as additional fire protection when the probability of bed ignition is high. Another way is to use a CO monitor. Low relative humidity (RH) increases the capacity of the carbon bed (because under high RH, the water is adsorbed and blinds the carbon). Manufacturers' recommendations on the maximum RH vary from as low as 40% to as high as 70%.

1.2.3 TYPES OF CARBON. Activated carbon used for vapor phase adsorption is different from that designed for liquid phase adsorption. Gas phase carbon has a larger number of small pores than liquid phase carbon.

1.2.4 ISOTHERMS. Isotherms for vapor phase adsorption of organic chemicals tend to be based more on calculated theoretical values, rather than on empirical data, which are limited. They are not as readily available in the literature as those for liquid phase adsorption. Isothermal data may vary greatly from one carbon series to another or among manufacturers. As a result, it is necessary to obtain vapor phase isothermal data from carbon manufacturers. The temperature and relative humidity of the vapor stream has a large effect on the adsorption capacity (the isotherms) of the activated carbon. Carbon adsorption increases as the temperature decreases. For example, lowering the temperature from 77 to 32°F at one site for one activated carbon resulted in increasing the adsorption capacity by 35%. High relative humidity can have a detrimental effect on the adsorption capacity. The difference in capacity from 0 to 100% relative humidity can be as much as a factor of 10. For example, increasing the relative humidity from 50 to 100% at an HTRW site decreased the adsorption rate from 0.12 g adsorbed per gram of carbon to 0.04 g adsorbed per gram of carbon. As a result, and since relative humidity depends on the temperature, it is often necessary to determine which combination of temperature and relative humidity is the most cost effective. Adjusting the relative humidity to 40 to 50% is often the best compromise. Relative humidity above 50 percent may result in adsorbed and condensed water vapor blocking the pores of the particles and interfering with the diffusion of the contaminants to the adsorption pores.

1.2.5 PRESSURE DROP. Headloss in vapor phase applications varies significantly, depending on piping configuration, carbon particle size, and surface loading rate. Surface loading rate is expressed as ft^3 (m^3) of vapor per ft^2 (m^2) of carbon bed cross-sectional area. For example, if the vapor flow rate is 100 ft^3/min (2.83 m^3/min.) and the cross-sectional area of the carbon bed is 10 ft^2 (0.929 m^2) (i.e., the diameter of the bed is 3.57 ft [1.09 m]), the surface loading rate is (100 ft^3/min)/10 ft^2 = 10 ft/min (3.05 m/min). Typical loading rates are 10–100 ft/min (3.05–30.5 m/min). A typical pressure drop through a vapor phase carbon bed is 1 to 4 in. of water column per foot of carbon bed (8.3 to 33 cm/m). In any case, the manufacturer's literature should be consulted regarding the headloss for a specific application.

1.2.6 OPERATING PARAMETERS. The major operating parameters needed to design a vapor phase carbon adsorption unit are:
- :Vapor stream flow rate.

- Contaminants to be adsorbed.

- Concentration of contaminants

- Temperature of the vapor stream.

- Relative humidity of the vapor stream

- Desired frequency between carbon bed changes

- Allowable pressure drop.

1.2.7 EQUIPMENT. The equipment and units needed in the adsorption process depend on the application. A typical process train consists of piping from the source of the volatile emission stream, such as vapor emissions from a soil vapor extraction unit, an induced draft blower, a heat exchanger, to raise or lower the temperature of the vapor stream (to adjust relative humidity), and carbon adsorption vessel or vessels.

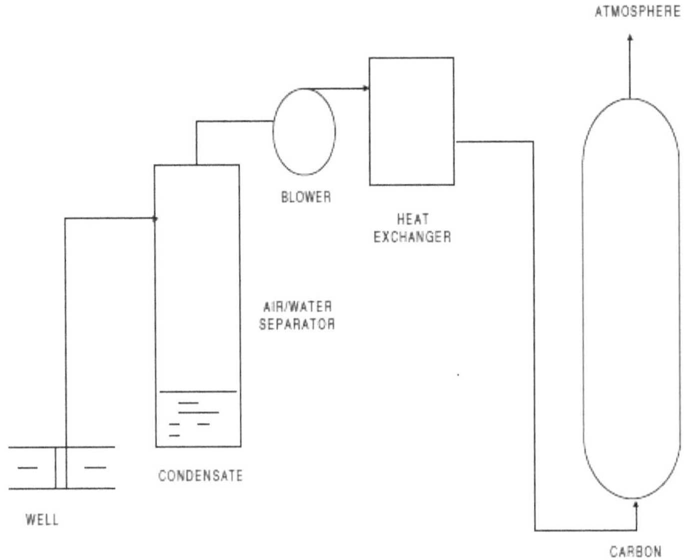

Figure 3-2.

Treating off-gas from an in-situ vapor extraction with activated carbon.

2. REGENERATION, REACTIVATION, AND DISPOSAL OF SPENT ACTIVATED CARBON.

2.1 ACTIVATED CARBON REGENERATION and Reactivation. This discussion presents information on reactivation and regeneration, options for spent activated carbon that has been used to treat hazardous wastes and industrial process effluents.

2.1.1 AS CONTAMINANTS ARE ADSORBED, the carbon's adsorptive capacity is gradually exhausted. When the carbon's adsorptive capacity is reached, it is considered "spent," and it must be regenerated, reactivated, or disposed of. Although some manufacturers and researchers use the terms "regeneration" and "reactivation" interchangeably, in this document, "regeneration" means removing the contaminants from the carbon without destroying them and "reactivation," which occurs at very high temperatures, means destroying the contaminants and reactivating the carbon. The user must decide which is to be used: on-site regeneration or reactivation, off-site reactivation, or disposal of the spent activated carbon.

2.1.2 REGENERATION USUALLY INVOLVES removing the adsorbed contaminants from the carbon using temperatures or processes that drive the contaminants from the carbon but that do not destroy the contaminants or the activated carbon. A common regeneration process introduces steam into the spent carbon bed, volatilizing the contaminants and restoring the carbon's capacity to what is called its "working capacity." Steam regeneration does not completely remove adsorbed contaminants. Another common process uses a hot inert gas, such as nitrogen, to remove the contaminants. The stripped volatiles are compressed, and recovered as liquid in a condenser. A third process is pressure swing adsorption. Pressure swing adsorption uses the fact that adsorption capacity is directly proportional to the partial pressure of the contaminants in the surrounding environment. The contaminants are adsorbed at a high pressure (providing higher partial pressure of the contaminant to be adsorbed), and then desorbed at a lower pressure where the capacity is reduced. These regeneration processes are usually run on-site and inside the adsorption vessel. All regeneration processes produce a waste stream that contains the desorbed contaminants. For example, steam regeneration produces a mixture of water and organics from the condensed desorbed vapor.

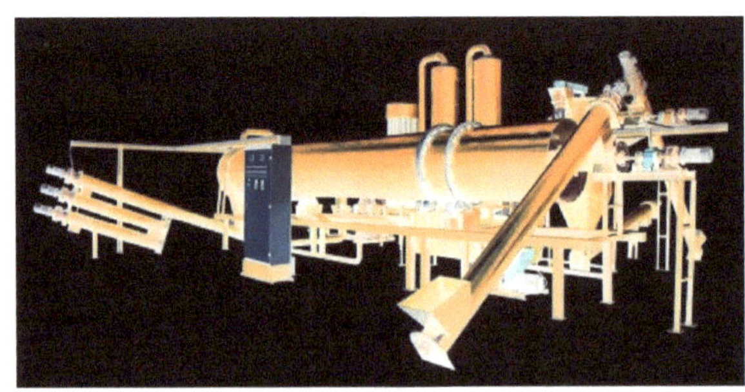

2.1.3 OTHER THAN THERMAL REACTIVATION at elevated temperatures, regeneration techniques will result in some contaminants remaining adsorbed and unaltered within the carbon particle. These contaminants will be occupying "high energy adsorption pores, or sites," and lower temperature regenerants (<500°F) or capacity corrections will not be able to provide sufficient energy to reverse the adsorptive force. Carbon having these residual contaminants remaining in the high energy adsorption sites will likely have much shorter runs before breakthrough. They might even be unable to attain the desired low-level effluent concentrations when placed back on-line, as compared to virgin grade carbons with all of their high-energy sites available for adsorption. These on-site regeneration techniques are based on capacity recovery processes traditionally used in solvent recovery operations and may not be suitable for applications driven by an effluent objective.

2.1.4 SPENT CARBON REACTIVATION OFF-SITE INVOLVES removing the adsorbed contaminants from the spent activated carbon in a process that is a modification of the one that initially activated the carbon. The contaminants are desorbed and destroyed in the high temperature (typically in excess of 1500°F [800°C]) pyrolizing atmosphere of the reactivation furnace. Several types of furnaces are available, such as rotary kilns and multiple hearths. The furnaces can be heated by a fuel such as natural gas or fuel oil or by electricity. Off-site carbon reactivation manufacturers reactivate spent carbon in large capacity (5 to 60 tons/day) furnaces. While furnaces of this capacity are not typically cost effective for a single hazardous waste site, smaller furnaces that may prove cost effective are available for on-site use from a number of manufacturers. Reactivation furnaces only produce reactivated carbon, air emissions, and some carbon fines. No organic wastes are produced. Table 3-2 summarizes the information for on-site regeneration, on-site reactivation, and off-site reactivation processes.

2.1.4.1A SELECTION CRITERIA FOR DETERMINING IF SPENT CARBON SHOULD BE DISPOSED OF, REGENERATED, OR REACTIVATED.

2.1.4.1 CRITERIA FOR DETERMINING WHEN TO USE ON-SITE REGENERATION, REACTIVATION OR OFF-SITE REACTIVATION, OR DISPOSAL.

2.1.4.1.1 ON-SITE REACTIVATION REQUIRES space and utility support for the equipment. It also usually requires an air pollution permit for the furnace afterburner. If the site cannot provide the land or utility support, or if obtaining the required permit is not practical, the spent carbon must be regenerated on-site or reactivated off-site.

2.1.4.1.2 AT SOME SITES, THE AVAILABILITY or turn-around times for off-site carbon re-supply may be impractical. In these situations, on-site regeneration or reactivation will be required or the site can provide sufficient storage for both fresh and spent carbon to eliminate the constraint of response time by outside suppliers.

2.1.4.1.3 STUDIES INDICATE THAT on-site thermal reactivation is not economical if carbon usage is less than 500 to 2000 lb/day (227 to 909 kg/day). Other studies have found that carbon reactivation unit cost rises rapidly if carbon usage is less than 5000 to 6000 lb/day (2272 to 2727 kg/day).

2.1.4.1.4 BECAUSE OF LIABILITY AND ECONOMIC concerns, some design guides recommend that reactivation should be done off-site whenever possible, regardless of whether land and utilities are available on-site.

2.1.4.1.5 TWO ALTERNATES, WHICH ARE VERY common today, are:

- Have a service come on-site and remove the spent carbon and replace it with virgin or reactivated. This operation usually takes less than one shift.

- Have an extra adsorber on hand and ship the adsorber with the spent carbon to a reactivator. The vessel will then be returned with virgin or reactivated carbon on it.

2.1.4.1.6 WHEN CARBON IS REGENERATED on-site, some contaminants may not be desorbed. For example, GAC containing organic contaminants with high boiling points may need to be reactivated instead of regenerated.

2.1.4.1.7 IN SOME SITUATIONS, THE LOSS of adsorption capacity or the rapid breakdown of the target contaminant, causing an unacceptable decrease in on-line time from the build-up of unregenerated contaminants, may be unacceptable. For these applications, reactivation (on- or off-site) will be required.

2.1.4.1.8 THE DESORBED MATERIAL PRODUCED by on-site regeneration processes may or may not be Resource Conservation Recovery Act (RCRA) wastes. In industrial applications, it may be possible to reuse or recycle the desorbed material. In hazardous, toxic, and radiological waste (HTRW) applications, the desorbed material is usually an unrecyclable mixture that requires proper disposal. Hazardous waste streams will have to be properly stored, manifested, transported, and disposed of. If it is not practical to handle a hazardous waste at the site, reactivation (either on-site or off-site) should be considered.

2.1.4.1.9 CARBON LOSSES DURING OFF-site reactivation in the adsorbers and the carbon transfer and handling systems can be held to 5 to 7%. Losses within the reactivation furnace should be between 1 and 5%, while potential total reactivation cycle loss rates can range from 3 to over 10%. Most systems operate with losses of 5 to 7%.

However, some researchers estimate that approximately 5 to 15% of the spent carbon is destroyed during each reactivation cycle. One manufacturer states that, in an on-site reactivation system, the losses range from 3 to 8%. The higher losses are experienced in locations that have a poorly designed carbon handling system, or where the adsorbed organics are difficult to reactivate or are strongly adsorbed on the carbon, or both. After the system shakedown is completed and the operators gain experience, carbon losses should decrease to approximately 7% per cycle. This loss can be replaced by carbon from the reactivated carbon pool or with virgin carbon. It is possible, although unlikely, that an inorganic contaminant in the replacement reactivated carbon might leach out at unacceptable levels in the effluent. Therefore, if the site must meet inorganic effluent limits, the operator should specify that virgin replacement carbon be used to make up the reactivation losses.

2.1.4.1.10 REGENERATION OR REACTIVATION will be necessary when no disposal site will accept the spent carbon or when the disposal costs would be prohibitive.

2.1.4.2 CRITERIA FOR DETERMINING WHEN TO DISPOSE OF SPENT CARBON.
There are several cases where regeneration or reactivation of the spent carbon will not
be feasible or will be prohibitively expensive. In these cases, the spent carbon must be
disposed of.

2.1.4.2.1 IF THE CARBON IS CONTAMINATED by a substance that damages it
irreversibly, it must be disposed of. For example, styrene monomer binds to carbon and
then reacts to form (polystyrene) polymers. These polymers blind the small micro pores
in the carbon and require extreme reactivation energies.

2.1.4.2.2 DISPOSAL MAY BE NECESSARY because regeneration/reactivation costs
may be prohibitive because of the site's location or because of trace contaminants, such
as radioactive particles, that are also adsorbed. Care should be taken to compare total
costs for both disposal and reactivation or regeneration. This comparison should
include: transportation costs to the disposal or off-site reactivation facility, the cost of the
continuing liability for the disposed of spent carbon, the continuing operations and
maintenance (O&M) costs for makeup carbon for on- and off-site reactivation, and the
capital and O&M costs for regeneration. The reactivators have trucks with three
compartments, so on-site storage tanks are not needed. In addition, different RCRA
regulations may apply to the spent carbon and to residuals from the on-site
regeneration or reactivation process. For example, the spent carbon, the slurry water
used to move spent carbon, and the water/contaminant mixture condensed from on-site
steam regeneration facilities may all be considered RCRA wastes because of the
"derived-from" rule (see 40 CFR 261.3(d)(1) for "derived from" rule for characteristic
waste and 40 CFR 261.3(2)(i) for listed waste). The costs of complying with the
appropriate storage, treatment, manifesting, and transportation regulations for these
wastes must also be included in the total cost comparison.

2.1.4.3 COMMON DESIGN CONCERNS FOR REGENERATION OF CARBON.
Because adsorption vessels and the spent carbon storage vessel will be pressurized or
put under vacuum, and in some cases heated (as with on-site steam regeneration), the

containers must be designed, fabricated, tested, and marked (or stamped) in accordance with the standards of the applicable Boiler and Pressure Vessel Code (ASME, 1992), and must incorporate pressure safeguards, such as rupture disks. Because wet activated carbon is corrosive, the vessel may be built with a corrosion allowance, typically 0.05 in. Most are protected with sprayed on linings, which range from 10 to 45 mils thick. An example of a coating used in carbon adsorption or storage vessels is 30 mils of vinyl ester. Other linings are fiberglass polyethylene, Teflon, and kynar. Once the coating is applied, it should be tested electronically to determine if there are any pinholes in its surface.

2.1.4.3.1 BECAUSE GRANULAR ACTIVATED carbon is abrasive, carbon loading and transfer piping and pumps should be built with an abrasion allowance. Carbon loading and unloading piping should avoid long runs, areas of low velocity, radical bends, and low spots without cleanouts, line restrictions, or restrictive bends. Another concern with piping is corrosion from the waste stream being treated. Chlorinated organics in vapor can corrode normal steels. Corrosion resistant materials such as Hastalloy or Alloy 20 may be considered.

2.1.4.3.2 WET, DRAINED, ACTIVATED CARBON adsorbs oxygen from the air. Therefore, all adsorption and storage vessels should include provisions to ventilate the vessels, and all inspection manways should be designed to support confined space entry procedures. In particular, the area around the manway should be designed to accommodate a rescue tripod. The inspection manways should also support the use of breathing air supplies, either as air supply lines or self-contained breathing apparatuses.

2.1.4.4 ON-SITE REGENERATION.

2.1.4.4.1 STEAM AND HOT INERT GAS REGENERATION. Steam and hot inert gas regeneration use the same principle. After the carbon bed reaches the end of its adsorption cycle, it is isolated from the contaminated waste stream. Steam or a hot inert gas (usually nitrogen) is piped into the adsorption vessel to strip the adsorbed contaminants from the carbon bed. The steam or gas can flow either counter-current or co-current to the original waste stream's flow. Currently, most systems use counter-current flow. The combined steam/contaminant or gas/contaminant is condensed and pumped to storage or treatment. Steam and hot inert gas increase the capital costs because more rugged materials are necessary construction and insulation.

2.1.4.4.1.1 STEAM/HOT GAS REGENERATION SYSTEMS are used primarily to regenerate vapor treatment beds, because the additional cost to dry out a water treatment carbon bed (raise temperature enough to vaporize all of the water entrained within the carbon pores) before regeneration makes steam/gas regeneration prohibitively costly. However, if the bed can be drained and dried before regeneration, steam/hot gas regeneration may be cost effective.

2.1.4.4.1.2 STEAM IS THE PREFERRED stripping gas, as it is readily available at many industrial sites; however, it may provide lower energy than hot inert gas, depending on the temperature. If it is not available, skid mounted boiler units are available at relatively low cost. Steam works especially well with non-water-miscible organics, such as chlorinated solvents. Non-miscible contaminants have an added advantage in that they can be separated from the condensed water by gravity. Steam is less useful for water-soluble contaminants such as alcohols, aldehydes, or ketones. If steam is used for these types of contaminants, the contaminants can be separated from the condensate by distillation. However, distillation raises the O&M costs of the system. For this reason, hot inert gas is preferred for water soluble contaminants.

2.1.4.4.1.3 THE REGENERATED BED IS COOLED, either by piping in cool air or water, or by simple radiation. Once the bed is cool, it is placed on standby or put into service as the polish unit. If a vapor adsorber was steam regenerated, the carbon bed must be dried before being put back into service. Conversely, water treatment units that are steam or gas regenerated must be carefully flooded after regeneration to remove any air or gas trapped in the carbon.

2.1.4.4.1.4 THE ADVANTAGES OF ON-SITE regeneration include the savings from not having to replace the 5 to 15% of the carbon destroyed during each reactivation cycle, no need for a carbon change out storage vessel, and the potential for recovery of the organic contaminants, with associated economic benefits. At some sites, primarily industrial sites, the recovered material is pure enough to be recycled. Also, the steam required for regeneration is already available at some sites and can often be supplied at minimal cost. Disadvantages include the need for storing the recovered contaminants, capital and O&M costs for a boiler if steam is not available, additional capital and O&M costs if hot inert gas is selected (for the gas, and for the condenser/chiller that will be needed), and the possibility that the system's carbon will have to be reactivated periodically anyway, owing to the buildup of contaminants that cannot be removed with steam or hot inert gas. At hazardous waste sites, there are two other potential disadvantages: the recovered material may be an un-recyclable mixture or steam condensate that must be properly disposed of, and the recovered material may be a RCRA waste, which must be stored, transported, and manifested according to RCRA regulations.

2.1.4.5 THE SYSTEM CONFIGURATION for steam regeneration must include a boiler, a feed water supply and treatment system, provisions for disposing of boiler blowdown, a condenser, a gravity separator, and storage for the recovered contaminants, either a tank or drums. If a mixture of petroleum chemicals and chlorinated solvents is being desorbed, the condensate may form three phases. This may complicate the disposal of the condensate. Vapor phase units require a source of drying air, such as process gas exiting an on-line adsorber or compressor. For hot inert gas systems, gas storage must be provided, either in cylinders or tanks, as well as a heater for the gas, a condenser, and contaminant storage. If economically feasible, an on-site gas generator may also be installed. Some systems use air as the stripping gas, avoiding the costs of gas storage. A separator is not usually required for gas systems because the condensate is a single organic phase. In addition, the carbon adsorbers must be plumbed for steam or the stripping gas, and piping to transfer the stripped contaminants to the condenser. A fan or pump for the cooling fluid will be needed if the units must be cooled down faster than radiation will permit. Most of these systems are not designed to be weather-proof and so should be located inside a building and protected from freezing. A typical process flow diagram for steam regeneration is shown on Figure 3-3.

2.1.4.6 IF A MANUFACTURER SUPPLIED the carbon regeneration system, they will supply the operation procedures and initial values for regeneration times and temperatures. Bench and pilot testing should also provide initial values for these parameters, especially the bed temperature that must be reached for effective regeneration. However, it must be emphasized that all of these parameters must be confirmed during start up and shakedown of the system. Even if bench or pilot testing was performed, the full scale system's initial values should be varied, because the full scale system's optimal settings will almost certainly vary to some degree from the optimal bench/pilot testing values. If bench and pilot testing were not performed on the actual waste stream, the initial operational system settings from the manufacturer should be conservatively modified (longer regeneration time and higher final regeneration temperature) until optimum parameters for the full-scale system are determined.

2.1.4.7 THE WASTE streams from most hazardous waste sites are not concentrated enough to generate sufficient heat to ignite the adsorption bed, but if a stream is very concentrated (e.g., soil vapor extraction vapor at the beginning of a remediation), the heat of adsorption should be monitored. (h) The reported time required to complete a regeneration cycle varied among the manufacturers contacted for this study because of a number of variables, including the contaminant load and the size of the regenerating equipment. One manufacturer advised that a 2000-lb carbon bed could be regenerated in approximately 3 hours using a 20-hp boiler.

2.1.4.8 IN-VESSEL STEAM AND HOT INERT GAS regeneration, even using superheated steam, will not reach the temperatures used by reactivation furnaces (at least 1500oF). Therefore, only compounds with boiling points less than the temperature reached in the vessel will be completely desorbed. This is not a serious problem if the contaminant stream is relatively pure, e.g., treating groundwater contaminated by a solvent spill. For a situation like this, the carbon bed can be regenerated by raising its final temperature above the boiling point of the sole contaminant of concern. However, when regenerating with steam, it is not necessary to have the temperature above the boiling point of the contaminants, as steam distillation occurs. To remove all of the contaminant higher temperatures are needed.

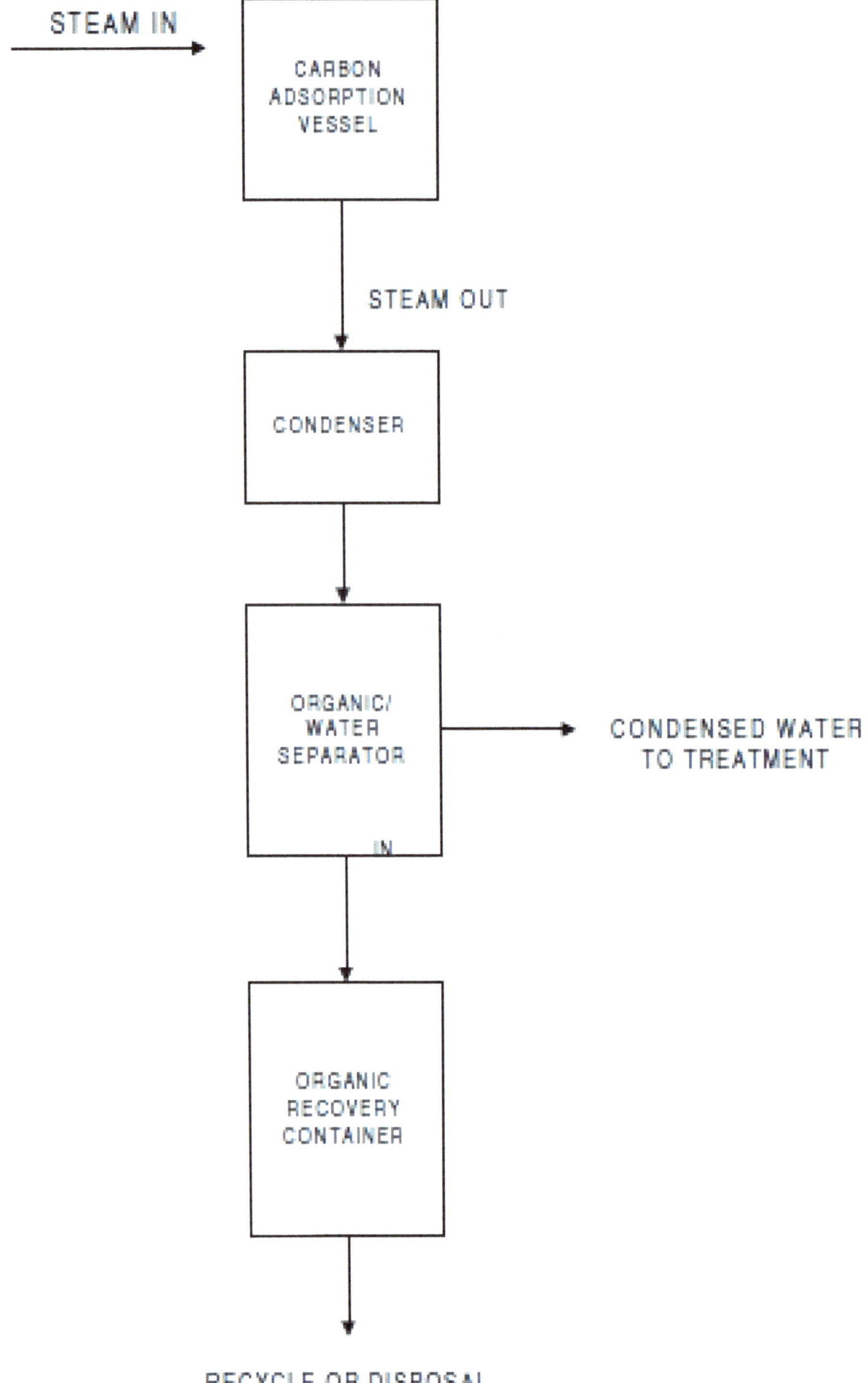

Figure 3-3

Steam regeneration

2.1.4.9 IF, HOWEVER, THE CONTAMINANT STREAM contains many contaminants, which is the norm for groundwater treatment operations at many hazardous waste sites, those contaminants with boiling points higher than the high temperature reached during steam or inert gas regeneration will not be removed. Because the system operator may not be analyzing for some of the contaminants that are not being removed during regeneration, the operator may not be aware that these compounds are fouling the carbon. Over time, these so-called "heavy boilers" can accumulate on the carbon and reduce its capacity. Because these compounds tend to have higher molecular weights, they clog the intermediate sized pores in the carbon, blocking access to the smaller pores that adsorb the lighter compounds of concern.

2.1.4.10 SPENT CARBON CAN BE TESTED to determine if regeneration is adequately removing the adsorbed contaminants. The overall capacity of the carbon can also be tested. (l) One method of removing these heavy boiler compounds is to send the carbon off-site periodically to be reactivated. Because steam regeneration will remove most of the contaminants of concern, off-site reactivation need not be done very often, perhaps no more than once per year. The actual off-site reactivation schedule will depend on the volume and composition of the waste stream being treated. Another removal method is to wash the carbon periodically with a solvent that will dissolve the heavier compounds. While this method avoids a carbon change out, it generates another waste stream that must be properly analyzed, stored, and manifested for off-site disposal or recycling. (m) The potential for this interference from heavy boilers can be investigated during bench or pilot testing if the actual waste stream to be treated is used to test the capacity of the carbon. The amount of the contaminants of concern adsorbed onto the test columns is routinely checked against the amount desorbed after each run. If the amount of contaminants of concern adsorbed decreases over several test runs, then it is possible that some unknown contaminant is accumulating on the carbon and decreasing its adsorptive capacity. If this interference is detected during bench or pilot testing, various solvents can be tested to determine their effectiveness in removing the contaminant. Periodic testing is recommended if the actual waste stream can change. Table 3-2 summarizes information for steam and hot gas regeneration systems and presents brief information on solvent and pressure swing regeneration not discussed further in this here.

Parameter	On-site Regeneration Steam	On-site Regeneration Hot Inert Gas	On-site Regeneration Solvent
Organic Chemicals that can be Desorbed	Most VOCs, Many SVOCs	Most VOCs, Many SVOCs	Any Organic that is Soluble in the Solvent
Carbon from Liquid or Vapor Systems Treated	Most Systems Treat Spent Carbon from Vapor Systems, Due to Extra Drying Costs of Liquid Systems.	Most Systems Treat Spent Carbon from Vapor Systems, Due to Extra Drying Costs of Liquid Systems.	Most Systems Treat Spent Carbon from Liquid Systems, Due to Extra Drying Costs of Vapor Systems.
Size Range or Process Rate	Systems Can Treat Up to 100,000 + cfm	Systems Can Treat Up to 100,000 + cfm	
Advantages	No Carbon Loss, compared to 5 to 15% of the Carbon Mass per Reactivation; Steam is Often Available at No or Low Cost; No Carbon Storage Vessel Needed; Potential for Recovery/Reuse of Contaminants.	No Carbon Loss, compared to 5 to 15% of the Carbon Mass per Reactivation; No Carbon Storage Vessel Needed; Potential for Recovery/Reuse of Contaminants.	No Carbon Loss, compared to 5 to 15% of the Carbon Mass per Reactivation; Ability to Regenerate Problem Contaminants, Such as "Heavy Boilers"; No Carbon Storage Vessel Needed; Potential for Recovery/Reuse of Contaminants.
Disadvantages	Condensate Storage must be provided; O&M and possibly Capital costs for the Boiler System; System Carbon may require Periodic Reactivation; and at Haz. Waste Sites, Condensate may be an Unusable Mixture that must be Disposed. Increased capital cost for materials of construction. Adsorbed organics remain on the carbon-capacity becomes "working capacity" which may cause short runs.	Condensate Storage must be provided; Capital and O&M costs for the Inert Gas System; System Carbon may require Periodic Reactivation; and at Haz. Waste Sites, Condensate may be an Unusable Mixture that must be Disposed. Increased capital cost for materials of construction. Adsorbed organics remain on the carbon-capacity becomes "working capacity" which may cause short runs.	Costs for Solvent Storage and Piping and Makeup Solvent, Problems with Recovering Water Miscible Solvents, Problems with Water Immiscible Solvents Penetrating All the Carbon's Pores, and Flammability Concerns with the Solvents.
Capital Costs (1)	Usually Less Than Reactivation Furnaces.	Usually Less Than Reactivation Furnaces.	Usually Less Than Reactivation Furnaces.
O&M Costs	Estimated at 1/3 of Reactivation. Costs for Steam Condensate Water Must Be Included.	Estimated to Be Approximately Equal To or Slightly More than Steam Regeneration.	Estimated to Be More Than Steam or Hot Inert Gas.

Table 3-2

On-site Regeneration, On-site Reactivation, and Off-site Reactivation Process Summary

Parameter	On-site Regeneration Steam	On-site Regeneration By Hot Inert Gas	On-site Regeneration Solvent
Residues to Manage	Water/Waste Mixture if Miscible, Organic Phase & Aqueous Phase if Non-Miscible	Waste Stream Condensate, or Air Emissions if Waste is Oxidized	Solvent/Waste Mixture
Manufacturers	AmCec; Continental Remediation Systems; Dedert; RaySolv; Wesport Environmental Systems; Vara International	RaySolv; Vara International	Waste Min, Design Only, PACS - Bench Scale Process, Only
Health/Safety	ASME Vessels Recommended, Wet Activated Carbon Adsorbs Oxygen, So Design Vessel for Confined Space Entry.	ASME Vessels Recommended, Include Ventilation for Vessel	ASME Vessels Recommended, Store Solvent and Solvent/Waste as per NFPA Standards.
EPA Regulations	Water (Steam Condensate) Treatment, Waste Stream Storage and Disposal or Recycle.	Condensate Stream Storage and Disposal or Recycle.	Water (Steam Condensate) Treatment, Waste Stream Storage and Disposal or Recycle
Pilot Testing Needed	Yes	Yes	Yes
Temperature Concerns	System Temperature Must Exceed COC's Boiling Point.	System Temperature Must Exceed COC's Boiling Point.	In Final Stage, System Temp. Must Exceed Solvent's Boiling Point.
Organic Chemicals that can be Desorbed	VOCs, Some VOCs	All	All

(1) Basis: 1996 Costs

Table 3-2 (continued)

On-site Regeneration, On-site Reactivation, and Off-site Reactivation Process

Summary

Parameter	On-site Regeneration By Pressure Swing Regeneration	On-site Reactivation Rotary Kiln	On-site Reactivation Multiple Hearth
Carbon from Liquid or Vapor Systems Treated	Treat Spent Carbon from Vapor Systems, Only. Usually on Storage Tanks or Processes.	Able to Treat Spent Carbon from Liquid or Vapor Waste Treatment Systems	Able to Treat Spent Carbon from Liquid or Vapor Waste Treatment Systems
Size Range or Process Rate	Systems Can Treat Contaminant Concentrations from 1,000 to 500,000 ppm.	Systems Can Reactivate 200 to 1,000 lbs per Hour.	Systems Can Reactivate 500 to 5,000 + lbs per Hour
Advantages	No Carbon Loss, compared to 5 to 15% of the Carbon Mass per Reactivation; No Carbon Storage Vessel Needed; Recovery of a Reusable Condensate Stream.	Complete Destruction of Contaminants; Control over Entire Process; Easier to Maintain than Multiple Hearth.	Complete Destruction of Contaminants; Control over Entire Process; More Fuel Efficient than Rotary Kiln; Better Reactivation Quality Control than Rotary Kiln.
Disadvantages	Higher Capital Costs than Other Regeneration Systems. Operationally Complex Systems. All Systems are Site-specific. Need to either adsorb at high pressures or desorb under vacuum conditions.	Loss of 5 to 15% of Carbon mass per Cycle (Average Loss -7%); Higher Capital Costs; Additional Space, Utility, and Training.	Loss of 5 to 15% of Carbon Mass per Cycle (Average Loss - 7%); Higher Capital Costs; Additional Space, Utility, and Training Requirements;
Capital Costs (1)	$500,000 to Several Million Dollars per System.	$150,000 to $700,000 +	$1,000,000 +/- for a 500 lb per Hour Unit.
O&M Costs	Requested But Not Provided by the Manufacturer	Reported at less than $0.05 per lb.	Costs Not Provided. Fuel Use estimated at 7 scf of Natural Gas per lb of Carbon.
Residues to Manage	Usually, Pure Product. Occasionally, Some Water Condensate.	Air Emissions, Only	Air Emissions, Only
Manufacturers	Radian; Design and Engineering Services, Only	College Research Corp.	Hankin Environmental Systems
Health/Safety	Requested But Not Provided by the Manufacturer	ASME Vessels Recommended. Wet Activated Carbon Adsorbs Oxygen, So Design Vessel for Confined Space Entry.	ASME Vessels Recommended. Wet Activated Carbon Adsorbs Oxygen, So Design Vessel for Confined Space Entry.
EPA Regulations	Condensate Stream Storage and Recycle.	Possible Air Emissions Requirements, Spent Carbon Storage Requirements.	Possible Air Emissions Requirements, Spent Carbon Storage Requirements.

(1) 1996 costs

Table 3-2 (continued)

On-site Regeneration, On-site Reactivation, and Off-site Reactivation Process Summary

Parameter	On-site Regeneration By Pressure Swing Regeneration	On-site Reactivation Rotary Kiln	On-site Reactivation Multiple Hearth
Pilot Testing Needed	Yes	Yes	Yes
Temperature Concerns	Heat of Adsorption must be Monitored and Managed by the System.	Reactivation Temperature Must be High Enough to Char and Gassify Contaminants, but Not So High that Excess Activated Carbon is Lost. Also, each Reactivation Stage's (such as Drying) Temperature Must be Controlled.	
Carbon from Liquid or Vapor Systems Treated	Able to Treat Spent Carbon from Liquid or Vapor Waste Treatment Systems	Able to Treat Spent Carbon from Liquid or Vapor Waste Treatment Systems	Able to Treat Spent Carbon from Liquid or Vapor Waste Treatment Systems
Size Range or Process Rate	System Can Reactivate 100 to 200 lbs. Per Hour.	Requested But Not Provided by the Manufacturer.	Limited by Manufacturer Site's Capacity, But up to 20,000 lbs. Of Carbon per Truck Load
Advantages	Complete Destruction of Contaminants; Control over Entire Process; System Footprint is Small; Low Utility Requirements.	Complete Destruction of Contaminants; Control Over Entire Process	Minimal Capital Costs, Especially if No Carbon Storage Vessel is Provided; Convenience, Complete Destruction of Contaminants: Quality Control of Reactivation Process Provided by Manufacturer.

Table 3-2 (continued)

On-site Regeneration, On-site Reactivation, and Off-site Reactivation Process Summary

Parameter	On-site Reactivation Electric Multiple Hearth	On-site Reactivation Electrically Heated Furnace	Off-site Reactivation
Disadvantages	Loss of 5 to 15% of Carbon Mass per Cycle (Average Loss – 7%; Higher Capital Costs; Additional Space, Utility, and Training Requirements. Need to either adsorb at high pressures or desorb under vacuum conditions. Corrosion control required on heater elements.	Loss of 5 to 15% of Carbon Mass per Cycle (Average Loss - 7%); Higher Capital Costs; Additional Space, Utility, and Training Requirements; Corrosion control required on heater elements.	Loss of 5 to 15% of Carbon Mass per Cycle (Average Loss - 7%); High System O&M Costs, Scheduling Concerns with Changeouts, Need to Provide Truck Access for Changeouts. Treatment site does not have permitting concerns. Off-site reactivators will often exchange "pound-for-pound" either virgin or reactivated, and site does not bear makeup requirements.
Advantages	Low volume of off-gas requiring treatment	Low volume of off-gas requiring treatment	
Capital Costs (1)	Manufacturer Prefers to Lease. If Purchased, Cost is $120,000 per Unit plus Royalty per lb of Carbon Reactivated.	Requested But Not Provided by the Manufacturer	Truck Access and possibly, costs for a Spent Carbon Storage Vessel.
O&M Costs	Costs Not Provided. Electricity Use Estimated at 1 to 2 kWh of Electricity per lb Of Carbon.	Costs Not Provided. Electricity Use Estimated at 1 kWh of Electricity per 1.5 to 2 lb Of Carbon.	Included in the Manufacturer's Reactivation Costs.
Residues to Manage	Air Emissions, Only.	Air Emissions, Only	Manufacturer's Responsibility
Manufacturers	COH Corp.	Custom Environmental International	Advanced Recovery Tech; Calgon; CETCO; Envirotrol; Nichem; Norit Americas; Service Tech; U.S. Filter/Westates
Health/Safety	ASME Vessels Recommended, Wet Activated Carbon Adsorbs Oxygen, So Design Vessel of Confined Space Entry	ASME Vessels Recommended, Wet Activated Carbon Adsorbs Oxygen, So Design Vessel for Confined Space Entry.	Changeouts Concerns, such as Vessel Pressurization, Change out Valve Sequencing, and Possibly Dust Control.
EPA Regulations	Possible Air Emissions Requirements, Spent Carbon Storage Requirements.	Possible Air Emissions Requirements, Spent Carbon Storage Requirements.	Possible Air Emissions Requirements, and Generator's Transportation Requirements.
Pilot Testing Needed	Yes	Yes	Yes

Parameter	On-site Reactivation By Electric Multiple Hearth Electrically Heated Furnace	Off-site Reactivation
Temperature Concerns	Reactivation Temperature Must be High Enough to Char and Gassify Contaminants, but Not So High that Excess Activated Carbon is Lost. Also, each Reactivation Stage's (such as Drying) Temperature Must be Controlled.	

(1) Basis: 1996 Costs

Table 3-2 (continued)

On-site Regeneration, On-site Reactivation, and Off-site Reactivation Process Summary

2.2 SAMPLING REQUIREMENTS. The spent activated carbon must be tested before a manufacturer will accept it for reactivation. Toxic characteristics leaching procedure (TCLP) analysis, total volatiles scan, PCB testing, and other testing may be required. Also, most off-site reactivators need to predict safe and satisfactory reactivation under their furnace operating parameters in order to accept spent carbons. One manufacturer (NORIT Americas) requires that each spent carbon have a Profile Sheet on file. To fill out this profile the facility operator would need to know if the spent carbon was a RCRA characteristic waste or a RCRA listed waste, the spent carbon's pH, and the Department of Transportation shipping name. In addition, the facility operator would have to be able to answer a number of specific questions, such as does the carbon contain Vinyl Chloride regulated under 29 CFR 1910.1017?

2.3 MANIFESTING, TRANSPORTATION, AND PLACARDING REQUIREMENTS.
When spent carbon meets the definition of a D.O.T. hazardous material (i.e., EPA hazardous waste), specific D.O.T. training requirements (49 CFR 172.700) will apply to persons shipping spent carbon off-site. The classification, management, and off-site disposition of spent carbon must be coordinated closely with the facility operator, and the installation. The facility operator will normally be responsible for the preparation of shipping papers, land disposal restriction notifications, etc. The installation personnel will normally sign the paperwork after it is prepared. Most carbon manufacturers can provide assistance to properly manifest the spent carbon. The carbon shipper is responsible for complying with all transportation and placarding requirements. If the material is a D.O.T. hazardous material, D.O.T. transportation requirements apply. The amount of insurance required per transport should be listed in the contract and reviewed periodically. Finally, the facility operator must be sure to obtain the appropriate Certificate of Destruction or Reactivation from the reactivation facility, listing how the adsorbed organic chemicals were destroyed or disposed of.

2.4 OFF-SITE CARBON REACTIVATION. These manufacturers are listed in Appendix D. Some manufacturers offer special services, such as segregating, reactivating, and maintaining a facility's carbon for that facility's own reuse. Makeup carbon for process losses can be obtained from the reactivated carbon pool or can be virgin carbon. Many users require that the makeup carbon be virgin. They do not want to take the chance of using carbon from other sources. However, it is very difficult to ensure complete segregation of small amounts of carbon, i.e., less than10,000 lb, as it moves through a large-scale reactivation process. Users need to verify that good quality control is employed. Off-site regeneration also has the advantage of the site receiving a known amount of constituent grade product and you need not be concerned over quality of product or amount of makeup carbon required. Table 3-2 summarizes the information on off-site reactivation processes.

2.5 NON-CARBON ADSORPTION.

2.5.1 GENERAL. Modified clay, polymeric adsorbents, and zeolite molecular sieves are also currently used in hazardous waste treatment. Some of these adsorption media are used primarily as pre-treatment for activated carbon. For example, these media may be used to remove compounds that may, through physical or chemical interactions, degrade the effectiveness of the activated carbon. Modified clay is primarily used as a pre-treatment for liquid phase systems, between an oil/water separator, and as a treatment process sensitive to emulsified oil, such as activated carbon or reverse osmosis. Without the use of the modified clay, the oils would blind the carbon, drastically lower its adsorption capacity, increase the cost of operations, and ultimately complicate the regeneration of the carbon.

2.5.1.1 THE ZEOLITES CAN ALSO replace activated carbon in several applications. Research indicates that the zeolites are mainly used for high volume vapor stream treatment. Table 3-3 summarizes the key differences among the media. Activated carbon is included for reference.

2.5.1.2 SOME OF THE POLYMERIC ADSORBENTS appear to be much more selective than activated carbon. For waste streams that have only one contaminant of concern, it may be possible to find an alternative adsorbent that is specific to the contaminant. By only adsorbing the single contaminant of concern, the working capacity of the alternative adsorbent may exceed the capacity of activated carbon, which loses some of its working capacity to competitive adsorption of other compounds in the waste stream. Also, for vapor phase applications, some of these alternative adsorbents are less affected by high relative humidity and high temperatures than activated carbon. For service in these environments, an alternative adsorbent may be able to provide treatment without pretreatment of the waste stream.

2.5.1.3 PRIMARY SELECTION CRITERIA FOR USING these alternative adsorbents and systems include the effectiveness of adsorbing the contaminants of concern and the overall lifecycle cost compared to using activated carbon. For most vapor or liquid service, both the proper alternative adsorbent and activated carbon will adequately adsorb the contaminants of concern. The alternative adsorbents usually have higher capital costs and lower operations and maintenance (O&M) costs. So, for short-term (2 years or less) projects, such as a one-time spill remediation, the alternative adsorbent will typically not be as cost effective as activated carbon systems. For long term projects, the lower O&M costs of the alternative adsorbent can make activated carbon less cost effective.

Parameter	Primary Treatment	Selective Adsorbent	Affected by High RH	Affected by Temperature	Reacts with Adsorbates	Media Capital Cost *	Media O&M Costs
Activated Carbon (AC)	Yes	No	Yes	Yes	Yes, especially Ketones	Low, $1.00 per lb	High
Modified Clays	No, Pre-treatment	No	NA, Liquid Treatment System	NA	No	Medium, $1.50 per lb (1)	NA, Once Through Material
Polymeric Resins	Yes	Can be Selective	Yes, but not as much as AC	Yes, but not as much as AC	No	High, $8 to $35 per lb	Low
Zeolites	Yes	Can be Selective	Not as much as AC or Resins	No	No	High, $7 to $10 per lb	Low

Parameter	Vapor Service	Liquid Service	Typical Adsorbents	Regeneration Methods	cfm/gpm Range	Health / Safety	Bench / Pilot Testing Needed
Activated Carbon (AC)	Yes	Yes	Nearly All Organics	Multiple; Steam, Pressure, and Solvent	to 200,000 + cfm, to MGD flows	Adsorbs oxygen, corrodes steel	Yes, but not as much as with Resins or Zeolites
Modified Clays	No	Yes	Emulsified Oil & Grease	NA, Once Through Material	2 to 200 gpm	Contains silica, control dust	Yes
Polymeric Resins	Yes	Yes	Chlorinated & Non-Chlorinated VOCs	Multiple; Steam, Solvent, Pressure, Microwave, and Hot Inert Gas	< 500 to 20,000 cfm 1 to 1,000 gpm		Yes
Zeolites	Yes	No	Chlorinated & Non-Chlorinated VOCs	Steam and Hot Air	7,500 to 200,000 cfm		Yes, but not as much as Resins

* - 1998 Costs

RH - Relative Humidity

NA - Not Applicable

NOTES: 1 - Rated medium because media can not be regenerated and reused.

Table 3-3

Alternative Adsorption Media Summary

2.5.1.4 A SUMMARY OF THE TECHNICAL INFORMATION needed to evaluate when alternative adsorption media may be selected in lieu of activated carbon is as follows:

- Media description.
- Adsorption system description.
- Availability.
- Estimated purchase and operating cost.
- Advantages and disadvantages for the application.
- Organic chemicals and contaminant ranges that can be adsorbed.
- Adsorption isotherms.
- Regeneration methods.
- Safety data and considerations, including loading, unloading, and handling methods.
- Applications.
- Pressure drop through the media.
- Effects of temperature and relative humidity.
- Any proprietary ownership and use limitations.

2.5.1.5 FOR GENERAL INFORMATION on isotherms, breakthrough, pressure drop, pilot tests, bed expansion regulations/disposal, safety concerns, pH (liquid phase), refer to these topics in the activated carbon paragraphs. Non-carbon adsorption media are very different from activated carbon. Design information must be obtained directly from the media manufacturer or the adsorption equipment supplier.

2.5.2 LIQUID PHASE NON-CARBON ADSORBENTS.

2.5.2.1 ORGANICALLY MODIFIED CLAYS. This material can be a mixture of anthracite and bentonite clay or bulk clay. The clay in both media has been treated with quaternary amine, which makes the surface of the clay much more active. One manufacturer, Biomin, Inc., reports that the clay removes emulsified oil and grease, and high molecular weight hydrocarbons, naphthalene, anthracene, COD, BOD, and heavy metals from liquid media. The material can remove up to 60% of its weight in oil and other organic chemicals; however, it cannot be regenerated. Disposal options include its being used as fuel if the oil adsorbed has sufficient heating value. Otherwise, disposal is governed by the type of organic contaminants the clay has adsorbed. Typically, the spent material has been incinerated, blended into cement kiln fuel, or treated biologically (e.g., landfarming), or placed in a landfill.

2.5.2.1.1 MODIFIED CLAY IS USUALLY APPLIED as a pre-treatment unit upstream of activated carbon or reverse osmosis units that might be blinded by emulsified oil. While the clay can remove large amounts of free oil, its capacity will be used up rapidly. Therefore, it is usually put on-line downstream of a gravity oil/water separator, so that the clay's capacity is used on emulsified oil only. Another application is as a final polish unit after an oil/water separator and before the treatment stream discharge.

2.5.2.1.2 GENERAL SPECIFICATIONS, design criteria (such as pressure drop through the system), recommended bed depth, hydraulic loading, recommended contact time, bed expansion during backwashing, swelling when wetted, and safety considerations must be obtained from the manufacturer. One manufacturer recommends a minimum bed depth of 3 ft, a hydraulic loading of 2 to 5 gpm/ft^2, a contact time of 15 minutes, and a headloss of 1 to 5 in. of water per foot of bed when contaminated.

2.5.2.1.3 NO PROPRIETARY OWNERSHIP or use limitations were mentioned in the manufacturers' data. The clay swells by at least 10% and sometimes as much as 20% when wetted, so the adsorption vessel should not be completely filled with the media. Also, the spent clay sticks together in "grapefruit" sized lumps, making it difficult to remove from the adsorber unless there is a side wall manhole. One manufacturer

recommends the clay media not be used in fiberglass vessels, as it can be difficult to
service these types of tanks once the clay is spent.

2.5.2.2 POLYMERIC RESIN ADSORBENTS. There are three primary types of
polymeric resins for liquid service: carbonized ion exchange resins, divinyl benzene
(DB) adsorbents, and post-crosslinked adsorbents. Table 3-4 compares several
features of these different media.

2.5.2.2.1 EACH OF THESE TYPES of resins is manufactured using a different process.
Because the manufacturers have greater control over the basic feed stock and
processing conditions, these materials can be "fine tuned" to a greater degree than can
activated carbon. For example, it is possible to create carbonized resins with pore
structures that will adsorb only contaminants of certain molecular weights. As with
carbon the EBCTs for liquid phase applications are typically much greater than EBCTs
for vapor phase applications.

2.5.2.2.2 PRESSURE DROPS THROUGH THE various media are usually included in
the manufacturer information. One manufacturer reports that pressure drops for liquid
phase systems ranged from psi/ft (23 kPa/m) of bed depth for a flow rate of 10 gpm/ft^2
to 10 psi/ft (226 kPa/m) for a flow rate for 100 gpm/ft^2. Pressure drop from one
manufacturer is a function of the velocity of the liquid through the adsorption bed raised
to a power (i.e., pressure drop = K1 (velocity) K 2). Manufacturers' literature often
represents this as a straight line on a log-log graph. Resins are patented by their
manufacturers. A license or other agreement is required to use them. Some resins can
be regenerated with steam, but specific information on regeneration must be obtained
from the manufacturer.

2.5.2.2.3 WATER TEMPERATURE IS NOT usually a problem for groundwater and
wastewater treatment. If the contaminated water's temperature is near a specific
contaminant's boiling point, however, the resin's desorption kinetics may be so fast that
desorption occurs almost as rapidly as adsorption. In this case, the resin's working

adsorption capacity will be too low to adequately adsorb the contaminant. For example, vinyl chloride is a gas at room temperature. Vinyl chloride- contaminated waters may have to be cooled, or extra capacity may need to be built into the adsorption bed, in order to get adequate adsorption. (d) Polymeric resins do not tolerate significant biological fouling. Polymeric resins can support biological growth, but the temperature of the steam regeneration (298°F) is usually sufficient to kill any biological film. All three primary types of polymeric resins, carbonized ion exchange resins, vinyl benzene (DB) adsorbents, and post-cross-linked adsorbents are also used for vapor adsorption. Table 3-4 compares several features of these different media. Table 3-5 lists the organic contaminants that can be adsorbed by three manufacturer's products.

2.5.2.3 ZEOLITE MOLECULAR SIEVES. ZEOLITE MOLECULAR SIEVES ARE NOT USED IN LIQUID APPLICATIONS.

2.5.3 VAPOR PHASE NON-CARBON ADSORBENTS.

2.5.3.1 POLYMERIC ADSORBENTS. Each type of phenolic resin is manufactured using a different process. Because the manufacturers have greater control over the feed stock and processing conditions, it is possible to create carbonized resins with pore structures that will adsorb only contaminants of certain molecular weights. Divinyl benzene adsorbents are hydrophobic, allowing them to be used in high relative humidity environments. These resins usually have very fast adsorption kinetics, which allows the empty bed contact time (EBCT) of the adsorber to be reduced. For example, a typical activated carbon EBCT for a vapor phase unit is 2 to 4 seconds. For a resin vapor phase adsorber, EBCTs can be as little as 0.02 seconds, allowing the designer to use much smaller adsorbent beds. The advantage of polymeric resins is that they do not react with the contaminants during adsorption to the degree that activated carbon does. For example, activated carbon adsorption is generally an exothermic reaction and there have been instances of activated carbon bed fires while treating highly contaminated streams. Polymeric adsorbents are much less reactive, allowing for fewer engineered safety controls on the system (Calgon, 1994).

2.5.3.1.1 PRESSURE DROPS FOR GAS THROUGH the various media must be obtained from the manufacturer. Typical pressure drops for a 40-cfm/ft2 vapor velocity ranged from approximately 4.5 in. of water per foot of bed depth to 45 in. of water per foot of bed depth.

2.5.3.1.2 ALL THE RESINS LOSE ADSORPTION CAPACITY as temperature increases. They also lose capacity as relative humidity increases, but at a lower rate than activated carbons do. As expected, the hydrophobic resins do better than the hydrophilic resins in high relative humidity service.

2.5.3.2 ZEOLITE MOLECULAR SIEVES. Zeolite molecular sieves are natural or man-made minerals composed of silicon and aluminum. These media have many of the same advantages as the polymeric resins. Zeolites are mainly used for high volume vapor stream treatment. The zeolites can be made hydrophobic, so they can be used in high RH environments. The engineered zeolites can be manufactured with uniform pore diameters, creating materials that selectively adsorb contaminants based on the contaminant's molecular size or weight. Because they are made entirely from inorganic oxides of silicon and aluminum, zeolites can withstand temperatures up to 800°C in dry air and up to 500°C in humid or steam environments. This temperature resistance allows zeolites to be regenerated at high temperature with air, eliminating the formation of contaminated condensate. Like the resins, zeolites are much less reactive than activated carbon when adsorbing ketones and other reactive organic chemicals. Also, zeolite's resistance to high temperatures allows the operator to burn off high boiling compounds or polymerized materials, like styrene, from the zeolite without damaging the media.

2.5.3.2.1 MANUFACTURERS CAN PRODUCE ZEOLITES especially for control of volatile organic chemicals that can adsorb the following organic compounds: benzene, toluene, xylene, phenol, cumene (isopropyl benzene), methylene chloride, trichloroethylene vinyl chloride monomer, alcohols, aldehydes, nitriles, aliphatics, CFCs, ketones, organic acids, and low molecular weight pesticides. Large molecules, such as multi-aromatic ring compounds, will not be adsorbed by zeolites because the molecules are too large to fit through the molecular sieve pore openings. These small pore sizes and the uniformity of pore sizes in the zeolites may prevent fouling by heavy boilers. The adsorption capacity of the zeolites is approximately 0.1 to 0.15 g of contaminants per gram of zeolite. Table 3-6 summarizes this information from UOP, a manufacturer of zeolites.

Parameter	Carbonized Ion Exchange Resin	Divinyl Benzene Resin	Post-Crosslinked Resin
Physical Shape	Spherical Beads	Spherical Beads	Spherical Beads
Surface Area	550 to 1100 (m^2/g)	400 to 700 m^2/g	900 to 1100 m^2/g
Sales Price *	$35/lb	$8 to $16/lb	$16 to $25/lb
Hydrophobic or Hydrophilic	Hydrophilic	Hydrophobic	Variable Hydrophobicity
Pore Size Distribution	Unique Pore Size Distributions	Larger Average Pore Sizes	Unique Pore Size Distributions
Crush Resistance	High, compared to AC	High, compared to AC	High, compared to AC
Reactivity of Resin Surface	Reactivity lower than AC	Reactivity lower than AC	Non-catalytic Adsorption Surface
*1998 Rates AC – Activated Carbon			

Table 3-4

Comparison of Polymeric Adsorbents

2.5.3.2.2 THE ZEOLITES ARE USED IN three different applications: pressure swing systems, temperature swing fixed bed systems, and temperature swing wheels. (c) Temperatures affect the adsorption capacity of the zeolites. Adsorption capacity fell from approximately 0.15 g/g 22°C to approximately 0.12 g/g at 60°C. However, as the media were designed to operate in very high temperature waste steam, temperature effects above 60°C are not expected to be significant. In most of the cases, the different RH adsorption curves are relatively similar, indicating no significant loss of capacity at high RH levels. Again, as the media have been engineered to be hydrophobic and organophilic, this was expected.

2.5.3.2.3 ORGANICALLY MODIFIED CLAYS. Organically modified clays are not used in vapor applications. d. Regeneration. Polymeric media can be regenerated at low temperatures, allowing nearly all systems to use on-site regeneration. Some activated carbon systems must ship the carbon offsite for reactivation. The polymeric media use a variety of regeneration methods, including hot nitrogen gas or air, microwave or infrared heating, and temperature-vacuum. These media are usually produced in the form of beads. The beads have high crush resistance, so attrition during loading and regeneration is usually much less than with activated carbon. While an activated carbon system can lose up to 12% of the carbon during each reactivation cycle, resin systems can operate at practically zero loss. For example, when American Society of Testing Materials (ASTM) Method D 5159, Standard Test Method for Dusting of Granular Activated Carbon, is used to test resins, the amounts of dust generated are so low as to be statistically insignificant.

Trade Name	OPTIPORE V493	OPTIPORE V502	OPTIPORE XUS 43565.01	OPTIPORE V323
Manufacturer	Dow	Dow	Dow	Dow
Media Type	Post-Cross Linked	Post-Cross Linked	Post-Cross Linked	Post-Cross Linked
Contaminant	Formaldehyde, MEK, Methanol, Terpene, Styrene, Toluene, Xylene, Acetone, Methanol, Isopropyl Alcohol, Butyl Acetate, Methylene Chloride, 1,1,1-TCE, TCA, and PCA	Toluene, Xylene, MEK, Acetone, Methanol, Isopropyl Alcohol, Butyl Acetate, Methylene Chloride, 1,1,1-TCE, TCA, and PCA	Xylene, MEK, MIBK, Acetone, Methanol, Isopropyl Alcohol, Butyl Acetate, Methylene Chloride, TCA, and PCA	Styrene

Trade Name	AMBERSORB 563	AMBERSORB 563*, 564, 572, & 575	AMBERSORB 600	Hypersol-Macronet Sorbent Resins **
Manufacturer	Rohm & Hass	Rohm & Hass	Rohm & Hass	Purolite
Media Type	Carbonaceous	Carbonaceous	Carbonaceous	Crosslinked Polystyrene
Contaminant	1,1,2-Trichloroethene and Chloroform	TCE, VC, MEK, Methanol, Cyclohexanoee, and Dichloromethane	TCE	Pesticides, Herbicides, Phenol, and Chlorinated Phenols

* - Ambersorb 563 was listed separately and grouped with Ambersorb 564, 572, and 575.

** - Purolite did not provide data matching specific resins to specific contaminants.

Table 3-5

Organic Contaminants Adsorbed by Polymeric Media

	Zeolite			
	HiSiv 1000	HiSiv 3000	HiSiv 4000	HiSiv 5000
Contaminants Adsorbed	Larger molecules, such as toluene and MIBK	Small molecules, such as acetone, ethanol, and methylene chloride	Larger molecules, such as isopropyl acetate and trichloroethylene	VOC mixtures, such as printing or paint-spray solvents
Application	Moderate concentrations with average humidity	High humidity applications	Low contaminant concentrations, high humidity	Low contaminant concentrations, high humidity
Regeneration	High temp. or reduced pressure, purge with air, steam, or other gasses	High temp. or reduced pressure, purge with air, steam, or other gasses	High temp. or reduced pressure, purge with air, steam, or other gasses	High temp. or reduced pressure, purge with air, steam, or other gasses
Physical Forms	Powder, Extrudate, 1/8 in. Tri-Lobe, ¼ in. Tri-Lobe	Powder, Extrudate, 1/8 in. Tri-Lobe, ¼ in. Tri-Lobe	Powder	1/8 in. Tri-Lobe

HiSiv zeolites are manufacturered by Universal Oil Products.

Table 3-6

HiSiv Zeolite Information Summary